Insects
A Compare and Contrast Book
by Aszya Summers

Many people think that you need to travel to far off places to see amazing animals. But did you know that some of the most diverse animals live in your own backyard?

Believe it or not, there are more insects than any other group of animals. In fact, between 70% and 90% of all animal species known to humans are insects.

Insects, like crabs and spiders, are arthropods. Unlike mammals, birds, fish, amphibians, or reptiles, arthropods don't have a backbone. They may have a hard outer shell (exoskeleton), a body separated into parts (segments) and pairs of legs that bend (jointed).

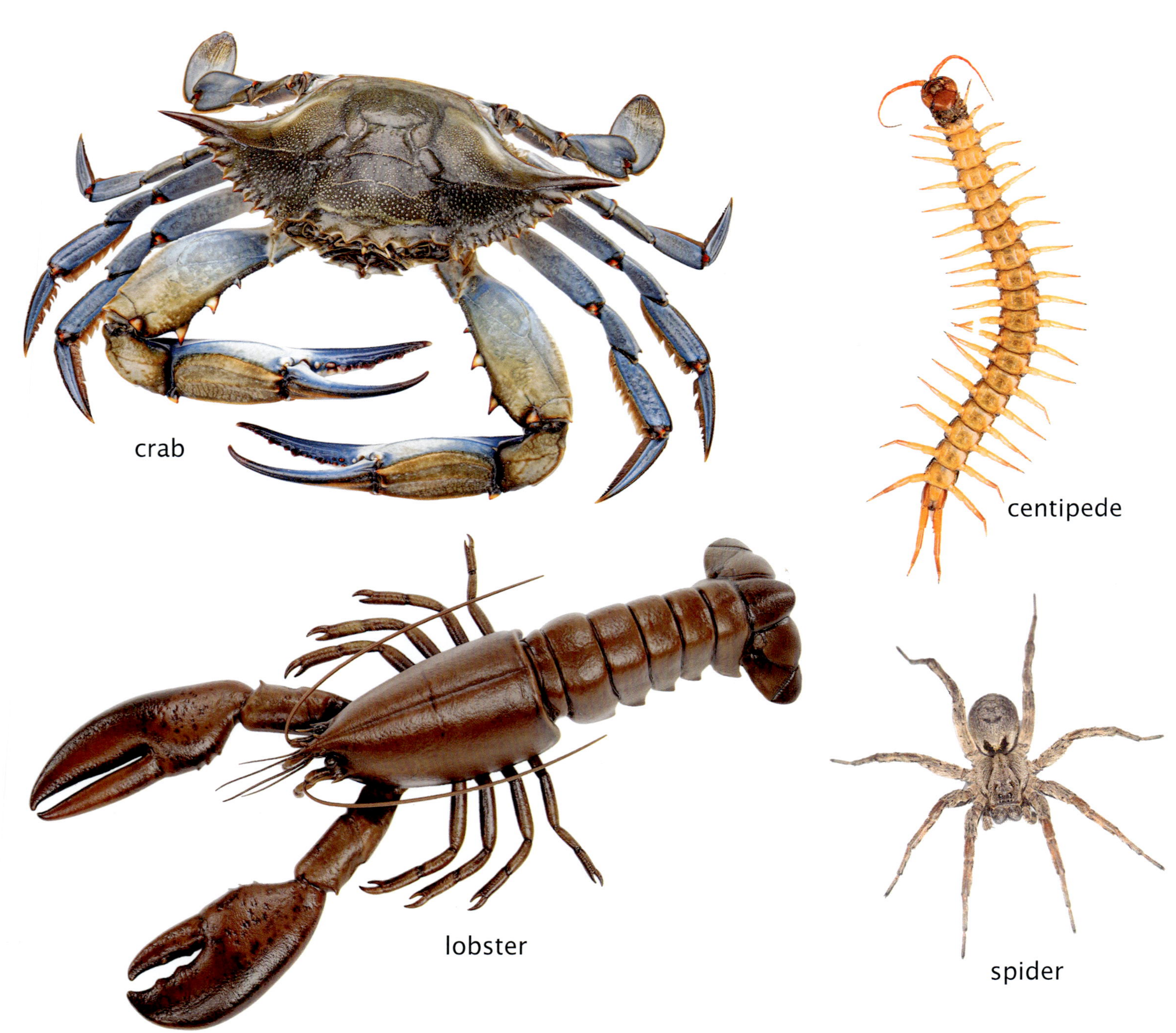

crab

centipede

lobster

spider

insects

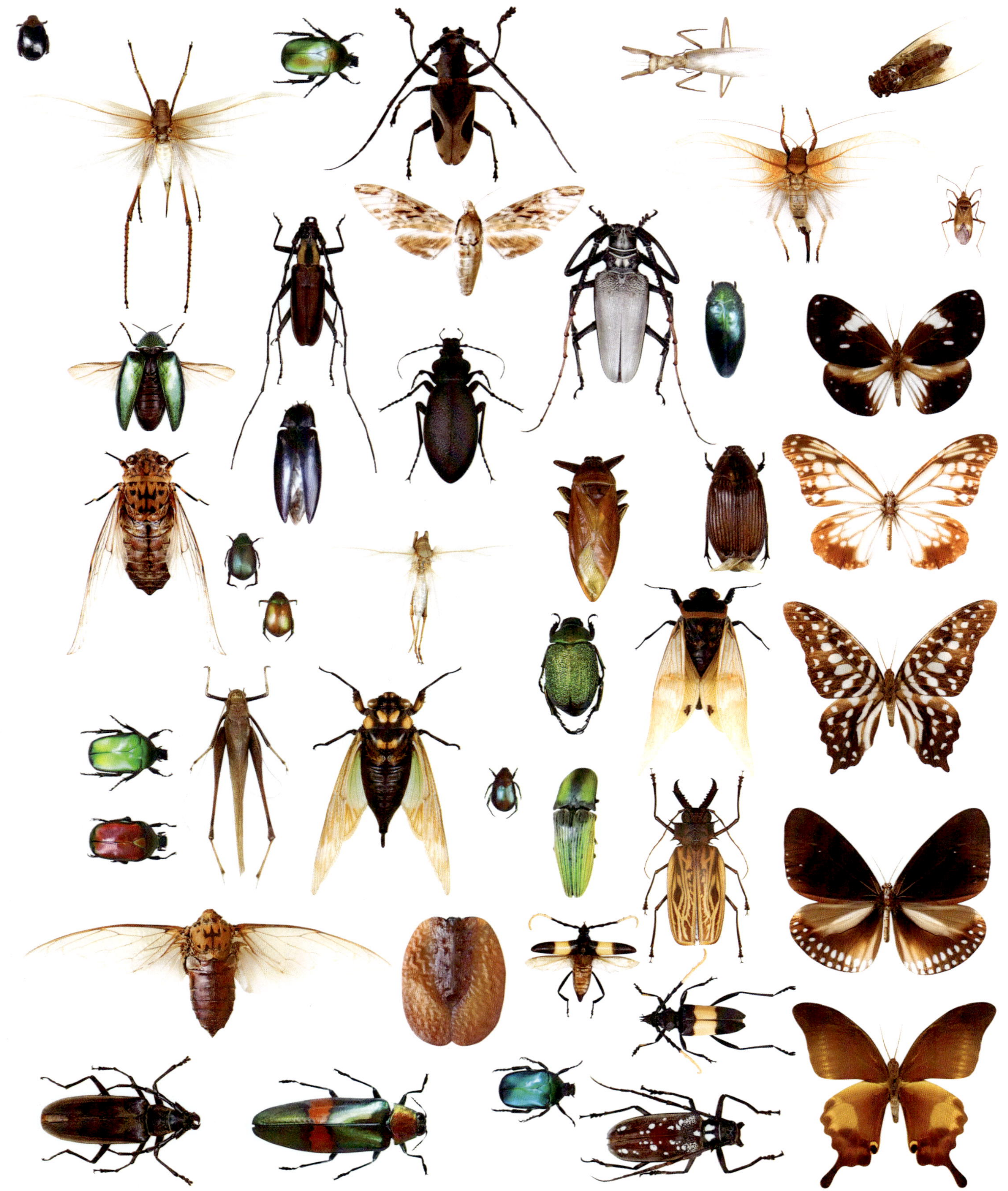

Unlike other arthropods, all insects have six legs and three segments to their body. Every insect, from butterflies to beetles, follows this body plan.

The abdomen has the insects' stomach and other organs inside.

Antenna and other sensing organs are part of the head.

Wings, if they have them, and legs attach to the thorax.

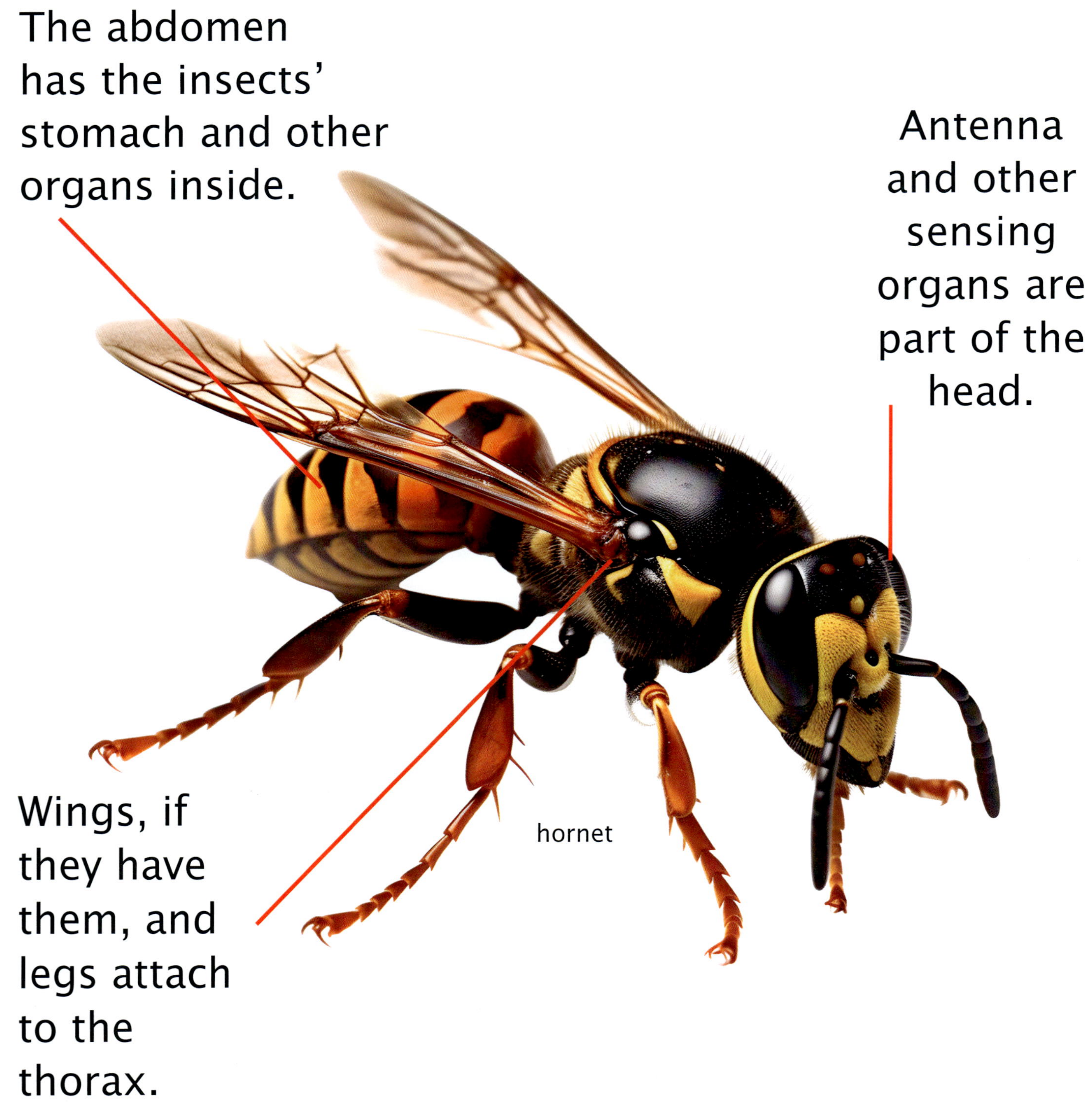

Can you find these body parts on these insects?

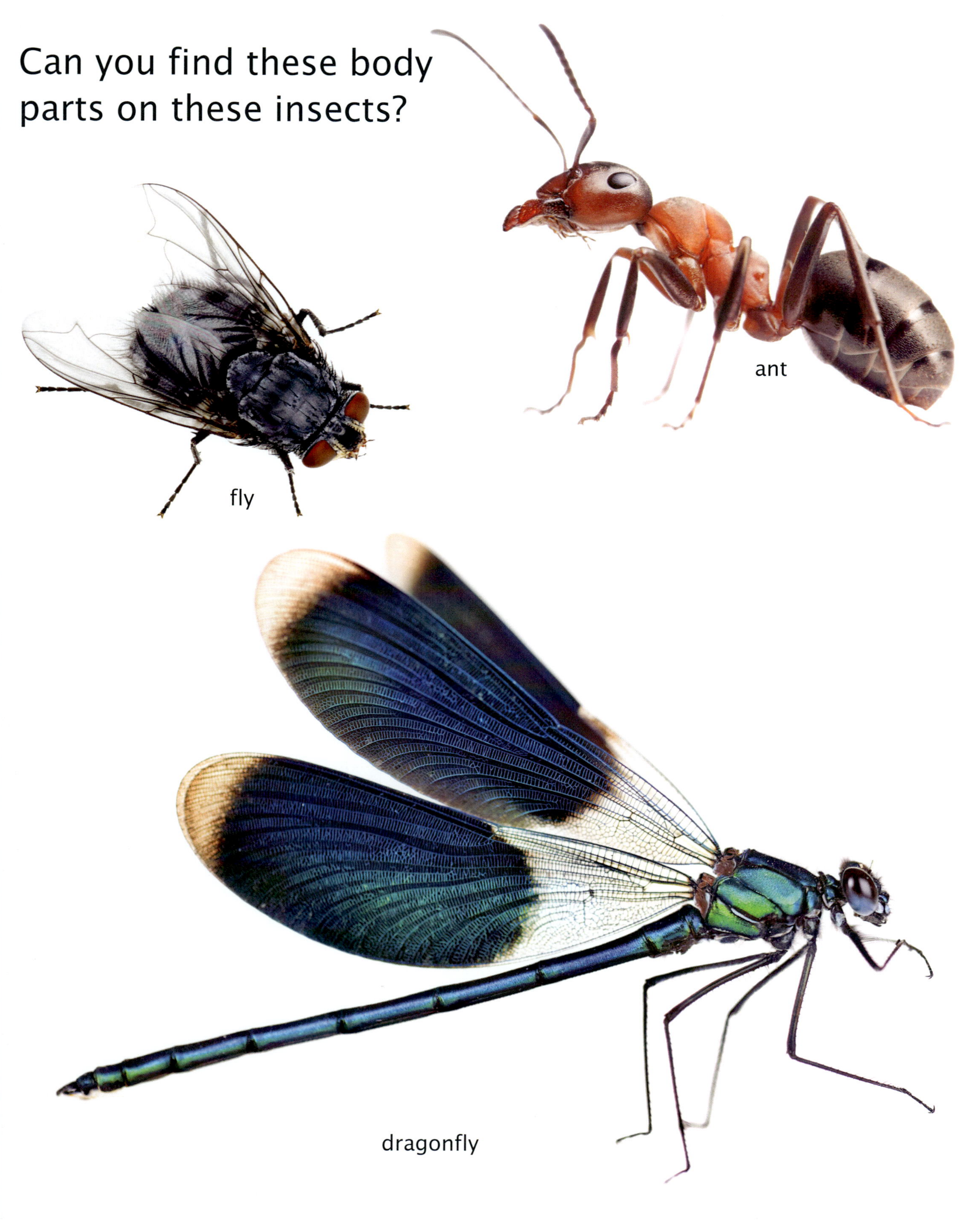

The largest group of insects are the beetles. Beetles have two sets of wings: a soft inner wing that folds up, and a stiff outer wing that keeps their delicate flight wing safe.

It's easy to see both pairs of wings on ladybugs.

Some beetles have crazy shapes, like this giraffe weevil. They use their long necks to roll leaves to protect their eggs.

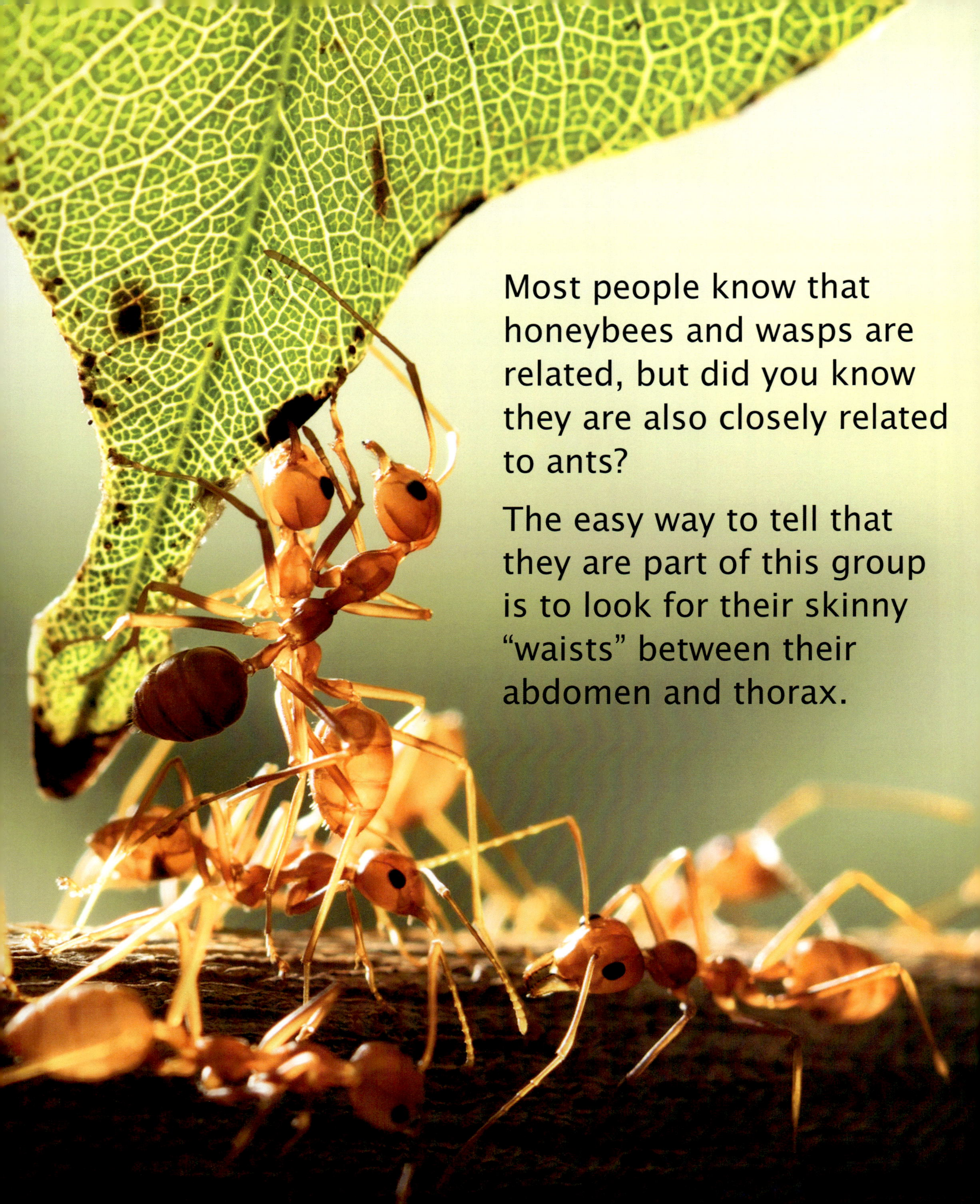

Most people know that honeybees and wasps are related, but did you know they are also closely related to ants?

The easy way to tell that they are part of this group is to look for their skinny "waists" between their abdomen and thorax.

Another trait that groups
these animals together is
that they make choices for
the good of their entire
community. A female queen
leads the group.

Butterflies and moths are well-known (and loved) insects.

Both have two pairs of large wings for flying, and a long, skinny mouth called a "proboscis" that is used for drinking liquid, like nectar from flowers.

In general, butterflies are brightly colored and come out during the day.

Moths generally come out at night and are drab in color.

But this is not always the case. This beautiful luna moth is common in the eastern U.S.

The best way to tell a butterfly from a moth is to get a close look at their antenna. Butterflies have long, clubbed antenna. Some have a small ball at the end. Moths have branching, comb-like antenna.

Another favorite insect group includes dragonflies and damselflies. Both hunt and eat other insects.

They both have two pairs of see-through wings.

Dragonflies have short, thick abdomens. Their back wings are usually fuller than their front wings. They rest with their wings open.

Damselflies have long, thin abdomens. Their wings are about the same shape and size. They sit with their wings closed.

Have you ever sat outside on a warm summer night and heard chirping from all around? You were probably hearing either crickets or grasshoppers. They both make that noise by rubbing their legs or wings together to communicate with others. They also use their strong, sharply bent legs for jumping.

Crickets are usually more active at night.

Grasshoppers are often seen
hopping around during the day.

Have you ever seen a stick walk? True to their name, walkingsticks camouflage by looking just like a stick.

Not all members of this group look like sticks, though. The walking leaf camouflages as a leaf instead of a stick.

Unlike plant-eating walkingstick insects, mantids are predators. Some have even been known to eat small birds or reptiles!

The most well-known mantid is the praying mantis.

Mantids are not left out of the camouflage game. While many are different shades of green, this orchid mantis mimics a flower to surprise and catch its prey.

fly
cockroach
flea

Some insects might even come into your home. Flies and cockroaches can be found in many homes. Fleas sometimes bother pets and make them itch.

While most insects are harmless, mosquitoes can carry diseases that may make humans sick. Fortunately, we can protect ourselves by using repellents.

As you can see, there is a lot of variety of wildlife all around us. Insects may be small, but they are mighty.

What is your favorite insect? What insects are you most excited to try to find and observe in your backyard?

For Creative Minds

Match the Insect "Cousins"

Using what you learned reading the book, match the insect "cousins" to each other. Insects shown are not to size.

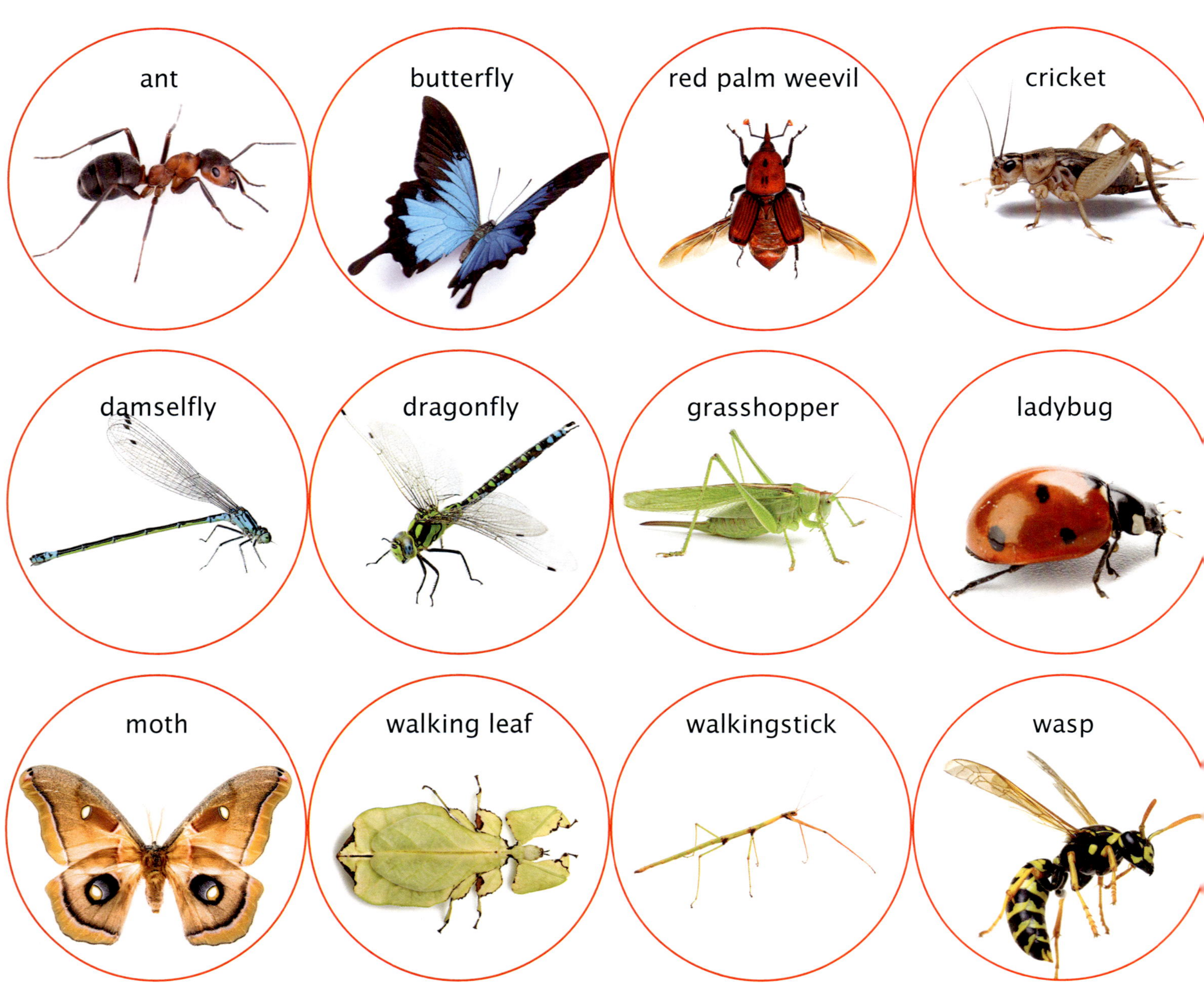

Answers: ant/wasp, red palm weevil/ladybug, butterfly/moth, cricket/grasshopper, damselfly/dragonfly, walking leaf/walkingstick

Insect Scavenger Hunt

They may be small and fast, but insects leave traces everywhere! Try to find as many different insect traces in your area as you can! Try looking at different times of day, or different seasons throughout the year!

This page may be photocopied or printed from **www.ArbordalePublishing.com** to track insect discoveries. Feel free to have kids track the dates, time of day, or even by season.

Find an anthill.	Find a leaf or flower that has been eaten.	Find some flowers with a bee or butterfly eating.	Listen for a cricket or grasshopper.

Find insects under a rock or log.	Find tiny holes in tree bark.	Find insects by digging in the dirt.	Find a mosquito bite.
			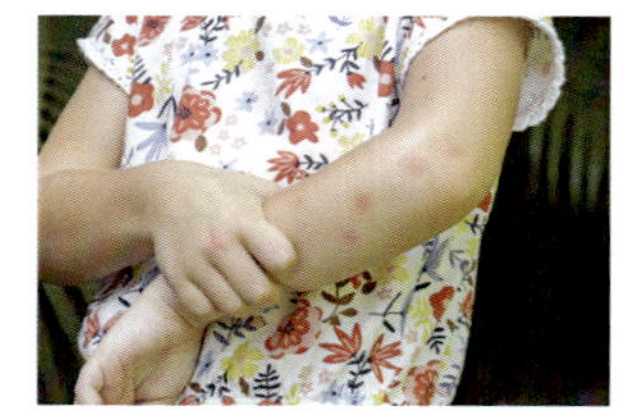
Find a cicada shell or hear one calling in the trees.	Find a ladybug.	Find a bee or wasp nest (but don't touch).	Shine a light outside at night to see what insects approach.

Insect Life Cycles

Depending on the type of insect, it either goes through a complete change, called metamorphosis or an incomplete or gradual metamorphosis.

- An Insect that goes through a complete metamorphosis begins its life when it hatches from an egg into a larva that doesn't look anything like the adult insect it will become.
- The larva eats and grows. In some cases, the larva adds body segments and goes through body changes (instars). As it goes through the instar levels, it molts its outer skin (exoskeleton) with a new one growing underneath. Whether a larva goes through instars or how many levels depends on the type of insect.
- At the end of the larval stage, the larva turns itself into a pupa.
- While inside the pupa, the insect's body is changing so that when it emerges from the pupa, it is a full-grown adult insect.

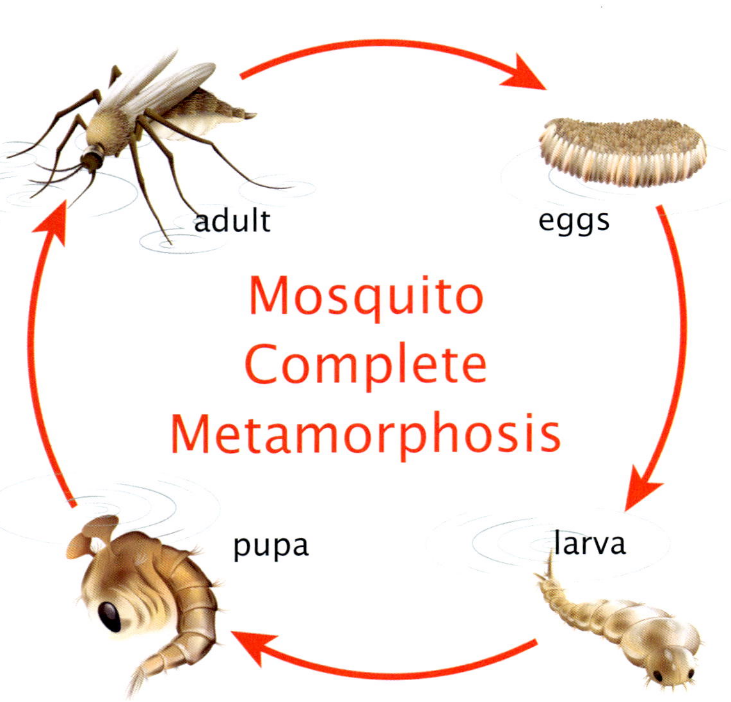

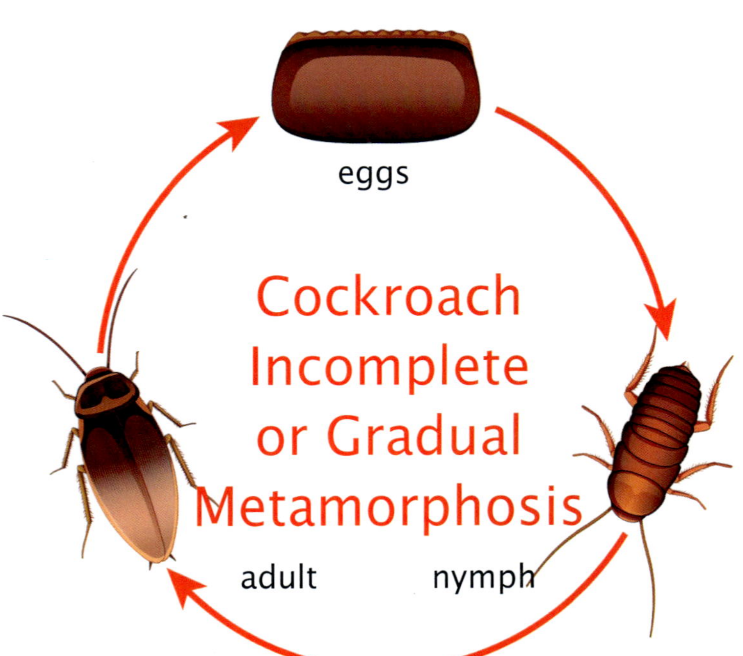

- An Insect that goes through an incomplete or gradual metamorphosis hatches from an egg into a nymph that looks like tiny versions of the adult insects but without wings.
- As the nymph grows, it molts its exoskeleton and grows a new, bigger one. A nymph will molt several times.
- By the time it has finished molting and growing, the insect has grown its wings and is an adult.

Complete or Incomplete Metamorphosis?

Based on what you learned on the previous page, see if you can determine whether the insect undergoes a complete or incomplete/gradual metamorphosis during its life cycle.

Larva is singular, larvae is plural.

atlas caterpillar (larva) molting

beetle grub (larva)

cicada nymph molt

cricket nymph molt

dragonfly nymph

spongy moth caterpillars (larvae)

swallowtail caterpillar (larva)

Did you know that a butterfly pupa is called a chrysalis
and a moth pupa is called a cocoon?

Answers: Complete: Atlas caterpillar/butterfly, beetle, swallowtail butterfly, spongy moth
Incomplete/Gradual: cicada, cricket, dragonfly

Thanks to Ian McAreavy, Lead Entomology Specialist at the Museum of Life and Science, for reviewing this book for accuracy.

All photographs are licensed through Adobe Stock Photos.

Library of Congress Cataloging-in-Publication Data

Names: Summers, Aszya, 1992- author.
Title: Insects : a compare and contrast book / by Aszya Summers.
Description: Mt. Pleasant, SC : Arbordale Publishing, LLC, [2023] | Series:
   Compare and contrast | Includes bibliographical references.
Identifiers: LCCN 2023054082 (print) | LCCN 2023054083 (ebook) | ISBN
   9781643519920 (English paperback) | ISBN 9781638170112 (Dual-language,
   read along) | ISBN 9781638170303 (pdf) | ISBN 9781638170495 (epub)
Subjects: LCSH: Insects--Juvenile literature.
Classification: LCC QL467.2 .S856 2023 (print) | LCC QL467.2 (ebook) |
   DDC 595.7--dc23/eng/20231213
LC record available at https://lccn.loc.gov/2023054082
LC ebook record available at https://lccn.loc.gov/2023054083

Also available in Spanish: ***Insectos: Un libro de comparaciones y contrastes***
Spanish Paperback   9781638173113
Spanish PDF   9781638173175
Spanish ePub3 9781638173205
The dual-language read-along is available online at www.fathomreads.com

Bibliography

"Dragonfly and Damselfly | San Diego Zoo Animals & Plants." Animals.sandiegozoo.org, animals.sandiegozoo.org/animals/dragonfly-and-damselfly.
Institution, Smithsonian. "BugInfo." Smithsonian Institution, www.si.edu/spotlight/buginfo.
National Geographic. "Spiders, Facts and Information." Animals, 29 Jan. 2019, www.nationalgeographic.com/animals/invertebrates/facts/spiders.